Sensors 2016

Topical Conference at the 2016 AIChE Annual Meeting

San Francisco, California, USA
13 - 18 November 2016

ISBN: 978-1-5108-3424-8

Printed from e-media with permission by:

Curran Associates, Inc.
57 Morehouse Lane
Red Hook, NY 12571

proceedings.com

Some format issues inherent in the e-media version may also appear in this print version.

Printed by Curran Associates, Inc. (2017)

For permission requests, please contact AIChE
at the address below.

AIChE
120 Wall Street, FL 23
New York, NY 10005-4020

Phone: (800) 242-4363
Fax: (203) 775-5177

www.aiche.org

Additional copies of this publication are available from:

Curran Associates, Inc.
57 Morehouse Lane
Red Hook, NY 12571 USA
Phone: 845-758-0400
Fax: 845-758-2633
Email: curran@proceedings.com
Web: www.proceedings.com

TABLE OF CONTENTS

460355 Label-Free Biomolecular Sensing Using Fluorescent Single Wall Carbon Nanotubes

Abstract:

A label-free molecular detection platform based on the fluorescent emissions of single wall carbon nanotubes will be presented. The carbon nanotubes are non-covalently modified with chelating polymers and single-stranded DNAs, which provides a specific recognition site for protein molecules. The bindings of the target biomolecules with the surface modified nanotubes disturb the dielectric environment of carbon nanotubes and hence cause their fluorescent emission property change. By monitoring their emission intensities, the specific recognition event on the nano scale is transduced to measurable macroscopic signals. In this presentation, we will demonstrate a miniature microarray platform to integrate multiple recognition sites onto a small area of glass slide. Using only a few microliter of the target sample, we are able to quickly detect multiple analytes using our nanosensor platform. The label-free detection system will be of great potential for product quality control and clinical diagnosing applications in the future.

471277 Multifunctional Nanoparticles for a SERS Portable Diagnostic Platform

Abstract:

Recent advances in hand held Raman instrumentation have enabled the realistic potential for point of care diagnostics based on Surfaced Enhanced Raman Spectroscopy (SERS). Magnetic based nanoparticle assays have been previously demonstrated to achieve selectivities and sensitivies superior to ELISA platforms for the detection of both viral capsid proteins and antibodies to West Nile Virus and Rift Valley Fever Virus in complex media. The development of gold coated magnetic nanoparticles further enhances sensitivity and enables rapid assays (<15 minutes) with reproducible results. Another 2 fold sensitivity was recently achieved through the use of antibody fragments in place of whole antibodies on the nanoparticle surfaces.

473503 Fabrication of Large Area Surface-Enhanced Raman Spectroscopy Substrate with Rough Surface PMMA Colloidal Particle

Abstract:

Surface enhanced Raman spectroscopy (SERS) has drawn significant attention over the past decades as a unique technique that can achieve detection at the single molecular level. However, cost-effective strategies for fabrication of the SERS substrates with large area uniformity and consistent signal enhancement are still illusive. Here we report an efficient fabrication route of SERS substrates via convective flow coating from micro-colloids, followed by thin film deposition of metals such as gold and silver. To enhance the surface plasmon resonance, we synthesize PMMA microspheres with rough surfaces, which increase the density of hot spots significantly compared with particles with smooth surface. The corrugated particles are synthesized with delayed addition of the crosslink agent (e.g., DVB) during dispersion polymerization. Key parameters that affect the surface roughness such as the reaction temperature, molecular weight of surfactant, the start time of delayed addition are studied systematically. The surface plasmon resonance of these substrates strongly depend on the degree of roughness on the support colloidal array, metal film thickness, and the type of meal thin film. The enhancement factor and uniformity of these SERS substrates for detecting crystal violet, are also been investigated.

465289 "Fabrication of Micropatterned Flexible Biofunctional Devices Based on Silk Proteins"

Abstract:

Polymers from nature provide exciting possibilities as biomimetic materials for the development of mechanically deformable and biodegradable systems. In particular, biopolymers from the silkworm - fibroin and sericin, are being extensively explored due to exceptional mechanical and optical properties, along with biocompatibility and degradability. Our group reported on a technique that permits the use of photolithography via photoreactive conjugates of silk proteins that behave like negative tone photoresists.[1] High resolution protein features can be precisely patterned at sub-microscale resolution (μm) at the bench-top over macroscale areas (cm), easily and repeatedly with high-throughput. This fabrication strategy using proteins can be used for application areas ranging from tissue engineering to bioelectronics, photonics, and drug delivery. We show how periodic, microstructured arrays can be patterned on flexible films to form structurally induced iridescent and functional, soft optical structures.[2] To demonstrate the broader applicability of silk protein lithography, we present the formation of a fully organic, biodegradable and flexible bioelectronic device. The conductive polymer poly(3,4-ethylenedioxythiophene):poly(styrene sulfonate) (PEDOT:PSS) with photo-sericin is used to form a photoreactive, aqueous-conductive ink. Conducting micropatterns are then formed on a flexible silk fibroin sheet using photolithography. We show how this flexible and conformable organic device can be used to sense electroactive biomolecules such as dopamine and ascorbic acid, or to encapsulate enzymes such as glucose oxidase for the specific detection of glucose.[3] The silk-based micropatternable functional composites are formed using all water based green fabrication approach and shown to be cell friendly and degradable. Such systems can find applications in implantable optical devices, in-vivo bio-sensors, and bio-optoelectronic devices.

References:

1. "Precise patterning of silk microstructures using photolithography" - NE Kurland, Tuli Dey, SC Kundu, VK Yadavalli, Advanced Materials, 25(43), 6207-6212, 2013.

2. "Biopatterning of silk proteins for soft micro-optics" - RK Pal, NE Kurland, C Wang, SC Kundu, VK Yadavalli, ACS Applied Materials & Interfaces, 7(16), pp 8809â€"8816, 2015

3. "Conducting polymer-silk biocomposites for flexible and biodegradable electrochemical sensors" - RK Pal, AA Farghaly, MM Collinson, SC Kundu, VK Yadavalli, Biosensors & Bioelectronics, 2016, in press

462037 Developing Low-Cost Nano/Micro Wire Sensors Based on a Seed-Mediated Solution Process

Abstract:

Nanowire sensors are widely recognized as the next generation building block for chemical sensing. They offer high commercial potential in many fields including defense, environmental monitoring, food safety, and industrial operations. Existing sensors tend to underperform on one or more of the sensor performance metrics, despite ongoing incremental improvements in readout time, miniaturization, and cost. Our low-cost nano/micro wire sensors are fabricated on microchips containing photolithographic patterns from solution droplets by applying an electrical pulse. It is a room temperature and reversible process allowing for portable and reusable products. The sensor performance is further tunable by versatile organic chemistry. Our technology is also compatible with both silicon and plastic electronics. The organic crystalline materials used for the sensing component are tetrathiafulvalene bromide charge transfer salt (TTFBr) and potassium tetracyanolantinate (KCP). The sensor is made by electrochemical deposition of TTFBr or KCP, from a solution droplet, on gold photolithographic patterns on glass. The sensitivity, specificity, and recyclability data are measured by impedance spectroscopy in the presence of various gases. The sensors can detect ammonia, hydrochloric acid, and other toxic or flammable gases at concentrations as low as 2 ppm. The specificity of the sensor corresponds to the polarity of the vapors. It responds more sensitively to more polar molecules. The sensor can be recycled at least 10 times without significant reduction in its performance. The likely mechanism is the interaction between the gas molecule and the organic nanowire crystalline face that leads to changes in bulk electrical conductivity of the nanowire. The conductivity can be sensitively tuned by the chemical nature of the nanowire.

463488 Preparation of Electrospun Nanocomposite Nanofibers of Polyaniline/Poly(methyl methacrylate) with Amino-Functionalized Graphene

Abstract:

Electrospinning technology has regained popularity in the last decade probably due in large part to the increased interest in nano-scale properties and technology. The electrospinning process is an efficient, relatively simple and low-cost technique to produce ultrafine fibers or fibrous structures of various polymers with diameters ranging from several nanometers to a few microns. Application of this process to electrically conductive polymers allows the production of conductive electrospun fiber mats that are very promising for many applications such as organic photovoltaics, scaffolds for tissue engineering and resistance-based sensors.

Gas sensors attract tremendous interest because of their widespread applications in industry, environmental monitoring, space exploration, biomedicine and pharmaceutics. Therefore, advances in sensor technology are constantly in demand, requiring improved sensors with faster response times, higher accuracies, increased sensitivities and other characteristics. Recently, 2D-graphene has attracted a lot of attention as a gas-sensing material since its discovery in 2004 due to its exceptional mechanical, thermal, chemical, electrical and optical properties. Moreover, it has high specific surface area that ensures every carbon atom as a surface atom and the electron transport through graphene may be highly sensitive to the adsorption/desorption of gas molecules. Graphene also has the ability to be dispersed in various polymer matrices thus creating a new class of polymer nanocomposites with new research dimensions in the fields of physics, chemistry, biotechnology and material science.

Among conducting polymers, polyaniline (PANI) remains the most popular due to its low cost, excellent optical and magnetic properties, high environmental stability and electrical properties that can be controlled by the oxidation and protonation state. However, it is relatively difficult to process compared to most other polymers. Polyaniline has a fairly rigid backbone due to its high aromaticity, and is commonly available only in relatively low molecular weight forms. Therefore, the elasticity of its solutions is insufficient for it to be electrospun directly into fibers. To circumvent this problem, it is possible to electrospun polyaniline by blending it with an easily electrospinnable polymer such as poly(methyl methacrylate) (PMMA). Moreover, polyaniline has successfully been demonstrated as an efficient gas sensor for monitoring organic and inorganic compounds.

In this work, we report the preparation of electrospun fiber mats of polyaniline doped with camphor-10-sulfonic acid, blended with poly(methyl methacrylate), and filled with amine-functionalized graphene using electrospinning for gas detection. The amine functionality in graphene will allow the high selectivity to carbon dioxide (CO_2). The combination of graphene with polyaniline using the electrospinning technique, as the conducting materials are good candidates for sensor matrices, with the advantages of low cost, easy preparation and ready modulation of sensing properties by varying their chemical structure. X-ray photoelectron spectroscopy, Fourier transforms infrared, Raman spectroscopy transmission electron microscopy and Scanning electron microscopy were used to characterize the amine-functionalized graphene and the morphology and properties of PANI/PMMA/AFG fiber mats, for detection of CO_2.

471637 Towards a Wearable Biochemical Sensing Platform: Hydrogel Enabled Osmotic Sweat Pumping for Microfluidics

Abstract:

The use of wearable devices in healthcare can greatly benefit from the development of new microfluidic sampling methods for sensing biomarkers in sweat. Therefore, we aim to develop a wearable platform for osmotic non-invasive microfluidic sweat sampling. By combining hydrogels with a microfluidic network, we seek to create a soft matter platform that is non-invasive and capable of pumping sweat from the body with no power requirement. The osmotic properties of hydrogels can draw fluid passively through a membrane and deliver it to a microfluidic network for biochemical sensing purposes. To demonstrate this concept, a microfluidic device was combined with thin biocompatible hydrogel discs equilibrated in saline or glycerol. The hydrogel interfaces with a water-permeable membrane, mimicking skin permeability. The high concentration of solute in the hydrogel creates an osmotic pressure difference across the membrane driving fluid flow through the membrane and into the microfluidic device. The release of this solute from the hydrogel autonomously pumps the fluid into an adjacent microfluidic channel. The osmotic strength of the hydrogel and its interfacial contact area with the membrane controls the flowrate. Continuous withdrawal of the fluid dilutes the concentration of solute in the hydrogel, which decreases the flowrate over time. Initial testing has shown that this sensing platform can pump biologically relevant and accurate levels of glucose across the membrane and through a microchannel to a reservoir interfacing a glucose sensor. Sensors and electrodes can be implemented into this microfluidic platform to test for other bioanalytes in sweat. This work represents an important step toward non-invasive sensing of sweat.

468169 Production of Hollow Carbon Microfibers upon Pyrolysis of Human Hair

Abstract:

Carbon is one of the most abundant materials in nature and has been widely used in diverse applications for many decades. Among various allotropes, such as diamond, graphite, coke, amorphous carbon, carbon nanotubes (CNTs), glassy carbon has received tremendous attention for various applications where its elegant physical properties such as high hardness, high electrical conductivity, low density, impermeability to gases, and high temperature resistance, are required. It has been widely applied as an electrode material for electrochemistry.

Herein, we present a novel method to fabricate glassy carbon microfibers from human hair wastes for electrochemical sensing. We demonstrate that the unique anatomy of human hair, composed of cuticle, cortex, and medulla, leads to long, hollow carbon structure upon pyrolysis at high temperature (900°C-1100°C) in an N_2atmosphere. At the elevated temperature, all the non-carbon elements are removed and only carbon remains.

The human hair, which we used as the carbon precursor, is an interesting waste material and has plenty of applications in industry and academic research. The coaxial structure of hair results in long hollow glassy carbon structures upon pyrolysis. The morphology of human hair samples before and after pyrolysis was characterized using scanning electron microscopy. The chemical composition of natural and pyrolyzed human hairs was also characterized using Raman spectroscopy and energy-dispersive X-ray spectroscopy. Screen printed carbon electrodes were modified with the hair-derived carbons and applied for electrochemical sensing of dopamine and ascorbic acid. The hair-derived carbons significantly improved the performance of the electrochemical sensors compared to the unmodified sensors. The presented method provides a simple, and inexpensive way to fabricate hollow carbon microfiber electrodes in various patterns.

470674 Fabricated Nanocomposite Temperature Detector Sensor

Abstract:

The aim of this work is to construct and test fabric like temperature detectors, which can be used for the advancement of exercise devices, prosthetic sockets, or medical applications. Resistance Temperature Detectors (RTDs) are an effective way to measure temperature due to the dependence of the RTD materialâ€™s resistance on temperature. The focus of this project is to develop RTDs for prosthetic socket or exercise applications, where the RTD is in close contact with the skin of the user, as in a prosthetic liner or in a show. Generally, RTDs are constructed from platinum, copper or nicker. However, metal RTDs have a tendency to be larger and rigid, making them uncomfortable for the user. The idea of this project comes from the need for a temperature device that does not create a pressure point within the liner or shoe. Rigid temperature detectors cause pressure points, which causes wear and sores on the user. The development of a fabric-like sensor would minimize such problems due to the thin, soft, flexible characteristics of the RTD. The fabric nanocomposite material acts like a RTD, because the conductivity increases linearly with temperature which is demonstrated by DC measurement. These are made from electrospun Nylon 6 and Polyurethane, which functionalized with multiwalled carbon nanotubes (MWCNTs) and polypyrrole (PPy) and characterized by SEM to show morphology of polymer matt/MWCNT/and polypyrrole. TGA was performed to shown the amount of MWCNTs adsorbed on the polymer substrate. The conductivity of device follows a percolate behavior and Temperature Coefficient of Resistance (TCR) which is experimentally determined. Resistance decreases with increasing temperature, which shows that this is a negative TCR material. The optimum TCR value is -0.204 ± 0.008%/C which shows that, this is suitable for temperature detection. This value is comparable to that of previously studied carbon nanotubes and devices, but uniqueness in that it is made using a polymeric substrate.

471052 Surface Acoustic Wave Sensors with Filled Microcavities and Waveguide

Abstract:

Surface acoustic wave (SAW) devices have been used for sensing for many years. The surface acoustic wave generated by the interdigital transducers (IDTs) of the device can be influenced by the mass loading on the device surface. When the molecules are bound on the surface, the velocity of the surface acoustic wave is changed, leading a shift in the phase or the frequency of the output signal. In this way, SAW devices can be used as the biosensors or chemical sensors.

To realize the potential of SAW devices in portable sensing applications, it is important to reduce the power consumption and improve sensitivity. In this work, we designed several SAW sensors based on ST 90°-X quartz. 3D finite element models (FEM) were constructed to model SAW devices with different structures, including bare device, device with a silicon dioxide (SiO_2) waveguide, device with microcavities in the delay path and device which incorporate both waveguide layer and microcavities. Then, the insertion loss (IL) and mass sensitivity of these SAW sensors were simulated to systematically evaluate the advantages of using waveguide and microcavities. We also compared the performances of two different materials (Tantalum and SiO_2) filled in the microcavities.

Our simulations demonstrated that the filled microcavities and waveguide layer can increase the displacements of the particles on the surface, and can also concentrate the acoustic energy to the surface of the device. As the acoustic wave propagates more closely to the device surface, less energy of the wave leaks to the interior of the device, which leads to a dramatic improvement in insertion loss. The SAW device becomes more sensitive to the mass loading on the surface because of the larger particle displacements and the concentration of energy. Therefore, the mass loading sensitivity of the SAW sensor is significantly increased by the filled microcavities and waveguide. The optimal thickness of the waveguide layer and the depth of the microcavities were obtained for the various devices and were found to be strongly dependent on the nature of the microcavity filling material.

Our FEM simulation results illustrated that a sensor device comprised of SiO_2 waveguide layer with Tantalum filled microcavities has the lowest insertion loss whereas a device configuration consisting of a SiO_2 waveguide layer with SiO_2 filled microcavities displays the highest sensitivity, compares well with our experimental observation. The SAW devices were fabricated on ST 90°-X quartz substrate. This study allows for construction of sensitive SAW biosensors, which work is underway.

455629 Frequency-Domain Approach to Determine the Separation Distance Between Addresses and Magnetic Readout Sensor in a Magnetoresistive Immunoassay Array Platform

Abstract:

Magnetoresistive (MR) sensors have emerged as a heterogeneous microarray-based assay platform for the rapid and sensitive detection of disease biomarkers labeled with superparamagnetic nanoparticles (SMNPs). Our platform consists of a periodic array of gold capture addresses and nickel reference addresses. Readout is performed in a manner similar to a hard disk drive by scanning the MR sensor across the array in a noncontact mode. Control of the separation distance between the MR sensor and capture surface is critical to detection because the magnetic field of the SMNPs decays with the cube root of the separation. That is, the separation distance should be minimized to maximize the readback signal; however, recognizing that the SMNPs have a nominal diameter ofÂ ~1 mm, a stand-off must be maintained to avoid contact with the SMNP label. To maintain the required stand-off, we employed the harmonic ratio method (HRM), which is used by the hard disk drive industry to maintain much smaller â€“ nanometer â€“ separations between a magnetic read head and moving data platter. This method relies on a comparison of the amplitude of the fundamental frequency of the readback response to those of its harmonics. When the signal is derived from a periodic magnetic array (*i.e.*, the Ni addresses of our array) that is scanned at a constant rate, the harmonic ratios contain the information necessary to determine and control the separation distance. In this presentation, we will demonstrate that the tenets governing the HRM used in the hard drive industry can be applied to our MR immunoassay array to determine separation distances with submicron resolution.

456380 Electrophoretic Cytometry: High Selectivity Single-Cell Protein Assays Enabled by Microfluidic Design

Abstract:

Dysregulation of signaling pathways is a hallmark of diseases including cancer. Cell-to-cell variation in this dysregulation is a critical aspect of disease development and progression. While single-cell resolution genomic and transcriptomic tools are advancing, these nucleic acid measurements do not directly measure protein-mediated signaling including dynamic post-translational protein modifications, protein localization, and protein-protein interactions. A majority of cancer therapies target proteins. Further, tumoral heterogeneity drives both drug response and resistance. Single-cell resolution protein tools can provide new insights. Powerful single-cell protein analysis tools do exist, yet all contemporary approaches to detect unmodified endogenous proteins are immunoassays (e.g., IHC/ICC, flow, and mass cytometry, and various ELISAs including microfluidic tools). Yet, the analytical specificity of immunoassays is dictated by immunoreagents (e.g., antibodies) which are notorious for sub-optimal performance. Limited availability of some immunoreagents severely curtails multiplexing capacity (e.g., off-target background signal) and the range of protein targets (especially isoforms). To mitigate these shortcomings of immunoassays, researchers prepend an electrophoretic protein separation to the immunoassay, thereby loosening constraints on immunoreagent performance and enhancing assay specificity. The tandem protein separation and immunoassay is known as a western blot. Unfortunately, the conventional western blot requires pooling of thousands of cells for each measurement, thus obscuring important cell-to-cell variation in protein expression and state.

We recently developed a microfluidic approach to achieve single-cell resolution western blotting, thus enabling direct measurement of proteins in single cells when high selectivity is required or when starting cell populations are exceedingly small. This presentation will discuss both fundamental and applied aspects of the targeted proteomics tools. In fundamentals, we will detail how a combination of separations science, microfluidics, and polymer formulation contributions offers a single-cell western blot with high throughput (>6,000 concurrent single-cell protein assays). In applied aspects, the presentation will highlight biological studies made possible by high selectivity protein measurements, including as related to drug resistance development in breast cancer, circulating tumor cells as indicators of cancer disease state, and as a tool to monitor sub-cellular localization of proteins as is relevant to signaling. Looking forward, we see single-cell resolution targeted proteomics as an important complement to nucleic acid measurements and discovery proteomics.

456529 Protective Polymer Coatings for the Enrichment of Rare Cells

Abstract:

Rare cells present a significant analytical challenge for sensing systems. An entirely new approach to cellular sorting will be presented that is based on the selective deposition of a protective coating on a cell and the destruction of all unprotected cells. Critically, the system is designed for high purity, where all uncoated cells are readily lysed. Purities >95% are common when sorting mixtures of cultured cells. Minority cancer cell populations are enriched from erythrocyte-depleted blood to >99% purity, whereas the entire batch process requires 1 h and <$2000 in equipment. Batch scale-up is only contingent on irradiation area for the coating photopolymerization, as surfactant-based lysis can be easily achieved on any scale. Multiple modes of protection will be discussed, as will their relationship to mechanical properties and molecular transport across the polymer coating. Finally, we will discuss the release of viable, functional cells from these protective polymer coatings.

452420 Cellular and Molecular Biosensing Using Magnetic Quantum Dots

Abstract:

Methods employing fluorescence for biosensing are well established using organic dyes. However, quantum dots offer several advantages to fluorescent dyes, most notably the potential for the detection of multiple biomarkers through multiplexed imaging, which is enhanced by their narrow emission bandwidth. Quantum dot biosensing can be further enhanced by combination with magnetic separation methods via the recent development of magnetic quantum dots. An advantage of magnetic quantum dot (MQDs)-based separation to current approaches is the small size of MQDs, which are small enough to potentially interact with single biomarkers/cell surface receptors, thus enabling relative quantization of results. MQDs can be independently manipulated by engineering magnetic fields, and can also be used to isolate molecules and cells for downstream analysis. Early stage clinical results will be reported using detection of circulating tumor cells via Her-2 expression in metastatic breast cancer patients as a model. Collectively, these results indicate a promising future for potential commercialization of biosensing technologies based on quantum dots and their magnetic quantum dot counterparts.

474657 Magneto-Nanosensors for Detecting Novel Protein Interactions and Early Stage Diseases

Abstract:

Substantial efforts have been made to understand the interactions between immune checkpoint receptors and their ligands targeted in immunotherapies against cancer. To carefully characterize the complete network of interactions involved and the binding affinities between their extracellular domains, an improved kinetic assay is needed to overcome limitations with surface plasmon resonance (SPR). Here we present a magneto-nanosensor platform integrated with a microfluidic chip that allows measurement of dissociation constants in the micromolar-range. High-density conjugation of magnetic nanoparticles with prey proteins allows multivalent receptor interactions with sensor-immobilized bait proteins, more closely mimicking natural receptor clustering on cells. The platform has advantages over traditional SPR in terms of insensitivity of signal responses to pH and salinity, less consumption of proteins, and better sensitivities. Using this platform, we characterized the binding affinities of the PD-1 â€" PD-L1/PD-L2 co-inhibitory receptor system, and discovered an unexpected interaction between the two known PD-1 ligands, PD-L1 and PD-L2.

The magneto-nanosensors are also well-suited for medical diagnostics of diseases at early stages because they possess attomolar to femtomolar sensitivities for protein biomarkers. The technology platform has been applied to in vitro diagnostics of lung cancer, prostate cancer, liver cancer, autoimmunity, radiation exposure, hepatitis, and cannabis abuse. The magneto-nanosensors are being adapted for point of care settings, mobile health, and global health, and Qualcomm Tricorder XPRIZE competition.

This work is supported in part by National Cancer Institute (NCI) through the Stanford Center for Cancer Nanotechnology Excellence (U54CA151459) and by National Institute of Allergy and Infectious Diseases (NIAID) through the Autoimmunity Center of Excellence (ACE) at Stanford (U19AI110491).

454408 Fully Integrated Wearable Biosensor Arrays for Multiplexed Sweat Analysis

Abstract:

Wearable sensor technologies play a significant role in realizing personalized medicine through continuously monitoring an individualâ€™s health state. To this end, human sweat is an excellent candidate for non-invasive monitoring as it contains physiologically rich information. Previously reported sweat-based biosensors either can only monitor a single analyte at a time or lack on-site signal processing and calibration mechanisms for accurate analysis of the physiological state. Given the complexity of sweat secretion, simultaneous and multiplexed screening of target biomarkers is critical and full system integration to ensure the accuracy of measurements is a necessity. Here, a mechanically flexible and fully-integrated perspiration analysis system is presented that simultaneously and selectively measures sweat metabolites and electrolytes electrochemically, as well as the skin temperature to calibrate the sensors' response. Our work bridges the technological gap between signal transduction, conditioning, processing and wireless transmission in wearable biosensors by merging plastic based sensors that interface with the skin with silicon integrated circuits consolidated on a flexible circuit board for complex signal processing. This application could not have been realized using either of these technologies alone owing to their respective inherent limitations. This wearable system is used to measure the detailed sweat profile of human subjects engaged in prolonged indoor and outdoor physical activities, and infer real-time assessment of the physiological state of the subjects. For example, the data obtained from outdoor running exercise shows that sweat sodium can potentially serve as an important biomarker for monitoring dehydration. The platform enables a wide range of personalized diagnostic and physiological monitoring applications.

472608 Improving Sensitivity of Electrochemical Sensors for Detecting Virulence Factors and Quorum Sensing Molecules in Pathogenic Bacteria

Abstract:

Improving sensitivity of electrochemical sensors for detecting virulence factors and quorum sensing molecules in pathogenic bacteria

Martin Kimani, Dr. Edgar Goluch, Department of Chemical Engineering, Northeastern University, Boston, MA

Microbiologists in the research and biotech industry still primarily depend on bacterial culture plates to identify bacterial species. In the clinical setting, a patient swab sample is submitted for identification which can take 1 â€“ 5 days before the clinicians can confirm a bacterial infection and provide the required antibiotic to the patient; this delay in acquiring results can hinder patient care especially with the growing impact of antibiotic resistant bacteria. Similarly during the production of biotech medications, bioreactor samples are submitted for offline analysis to monitor potential bacterial contaminations with results received anywhere from 1 â€“ 5 days after sample submission; this delay in acquiring bacterial contamination results can hinder production and negatively impact product supply to the patients. There are existing bacterial identification methods that reduce the time to acquire results from days to hours with improved sensitivity such as genetic identification using polymerase chain reaction (PCR) and other molecular methods that detect DNA hybridization. However, these technologies require costly equipment and therefore are not economically scalable for portable bacteria detection or for in-process monitoring of bioreactor contamination.

Our proposed technology provides an alternative low cost and real time bacterial detection to improve patient care and potential to provide in-line bacterial process monitoring. The technology uses low cost electrochemical sensors for detecting the redox active bacterial or molecular markers that bacteria excrete as virulence factors and for quorum sensing (QS). QS is a form of inter-cellular communication that bacterial species use to regulate gene expression and to synchronize multicellular activity. Our lab has previously used electrochemical sensors to detect one of the redox active virulence factors named pyocyanin that is produced by the clinically relevant bacteria *Pseudomonas aeruginosa*. The concentration of the pyocyanin was correlated to be proportional to the bacteria density. Therefore, the sensitivity of the sensors are limited to the relevant cell density.

In this work we explore how the sensitivity of the electrochemical sensors can be improved for detecting virulence factors and quorum sensing molecules in extremely low cell density states as those found in early bacterial infection and early bioreactor contamination cases. Tangential flow filtration coupled with forward osmosis was utilized to concentrate different species of bacteria cultures and redox active molecules in a flow reactor. Low cost electrochemical sensors were developed and used for in-process monitoring of the electrochemical activity of the concentrated cultures and redox active molecules. This work explores if and to what limits a low electrochemical signal from low concentrations of redox active molecules in a sample volume can be increased by forward osmosis.

The results from this work could provide the ability to detect virulence factors and quorum sensing molecules in extremely low concentrations similar to the concentrations that may be found during the onset of bacterial contamination. In addition, the improved electrochemical sensitivity from this work can allow for screening other bacterial species to identify if they produce extremely low concentrations of quorum sensing molecules and virulence factors that could be detected using electrochemical sensors.

469701 DNA-Based Sensor Particles Enable Distributed Light and Temperature Measurement in Difficult to Access Areas

Abstract:

With continuous technology miniaturization it is becoming increasingly important to measure properties and characterize systems on a micro- and nanoscale. The demand to assess previously inaccessible setups, such as underground reservoirs or single cells, has spurred the development of nanosensors – devices that transduce a chemical, physical or biological signal from a nanoparticle to the macroscopic world. Currently used nanosensors produce either an optical, mechanical, electrical or magnetic signal in response to a property that is measured. Such information is not stored over time inside the sensor and has to be read out immediately. If a measurement is done inside an organism, underground, or in closed setups, on-line signal transfer can be challenging or even impossible.

In this contribution, we present submicron size particles capable of measuring physical properties in remote areas and storing the recorded information until a readout. One attractive way to chemically store information is inscribing it into a DNA sequence, which can later be read out. [1, 2] However, for a physical property (e.g. light or temperature) to be saved in nucleic acids, the chemical structure of DNA has to be changed as a result of a physical stress that a sensor has been exposed to. We developed a sensing platform based on encapsulating two distinct species of nucleic acids inside silica particles (~150 nm) and using their stability differences to measure physical properties.

To quantify temperature, we used nucleic acids with different degradation kinetics (DNA and RNA) and related the difference in their decay (measured by quantitative polymerase chain reaction (qPCR)) to temperatures that a sensor has been exposed to. Such nanothermometer allowed measuring temperatures between 55° and 80° C. The scale could be further tuned by tailoring the nucleic acid sequences towards more specific applications. [3] To measure the amount of light irradiation, ultraviolet-sensitive protecting groups (caging groups) were installed on a sensing DNA molecule. As soon as light reaches a particle

containing caged DNA, the protecting groups are gradually released and the amount of uncaged DNA is quantified using qPCR and related to the cumulative irradiation. As a proof of concept, we measured light exposure in paramecia, a single-cellular organism. With a detection limit of individual cells, we could distinguish between cells that have been exposed to different duration of sunlight (between 0 and 2 hours). [4]

An important attribute of the DNA-based chemical sensors is the information storage capacity: in addition to the cumulative temperature or light information, nucleic acids serve as an identity unit, enabling tracing its origin. Therefore, such DNA-based tracer/sensor particles can be applied to track the flows and measure properties in environmental monitoring, material quality control, reservoir characterization, or other complex settings.

[1] Grass, R. N., Heckel, R., Puddu, M., Paunescu, D. and Stark, W. J. (2015), Robust Chemical Preservation of Digital Information on DNA in Silica with Error-Correcting Codes. Angew. Chem. Int. Ed., 54: 2552–2555.

[2] Paunescu, D., Fuhrer, R. and Grass, R. N. (2013), Protection and Deprotection of DNA—High-Temperature Stability of Nucleic Acid Barcodes for Polymer Labeling . Angew. Chem. Int. Ed., 52: 4269–4272.

[3] Puddu, M., Mikutis, G., Stark, W. J. and Grass, R. N. (2016), Submicrometer-Sized Thermometer Particles Exploiting Selective Nucleic Acid Stability. Small, 12: 452–456.

[4] Mikutis, G., Mora, C. A., Puddu, M., Paunescu, D., Grass, R. N. and Stark, W. J. (2016), DNA-Based Sensor Particles Enable Measuring Light Intensity in Single Cells. Adv. Mater., 28: 2765–2770.

472844 Single Electron Device As an Electrochemical Biosensor to Measure Biochemical Activity in a Cell

Abstract:

The conductance of nanoparticles modulates on charging by a single electron making them attractive for use in highly sensitive sensing devices. However, devices made from a single nanoparticle or arrays of nanoparticles have to be cooled to cryogenic temperatures to obtain the single-electron effect. Recently, we demonstrated the single-electron effect at room temperature in a monolayer network of one-dimensional "necklaces" of 10 nm Au particles. The necklace array network architecture is a metamaterial where the band gap can be tuned from 0 to up to ~6 V at room temperature by regulating the array's overall shape. The high conductance band gap of over 1 V at room temperature allowed a highly sensitive electrochemical field-effect transistor (eFET) to be fabricated that operates in salt solutions. Biofunctional activities were monitored in real time by interfacing the eFET to living cells. In one application, the infection of a single virus on an algal cell immobilized on an eFET was measured in real time. In another application, the membrane potential modulation during photosynthesis was measured in real time as a function of the carbon source in the solution. The single-electron effect in a necklace array in air, the eFET performance, and the application as a cell-based biosensor will be presented.

471295 A Zeolite-Assisted Microwave Planar Resonator Sensor for Non-Contact Quantification of CO_2 and CH_4 in Gas Streams

Abstract:

Gas sensing and analysis by monitoring the dielectric properties of the adsorbent materials is an emerging, robust, sensitive, and inexpensive technique with contactless operation capability. Planar resonator sensors are particularly favored for this purpose due to their simple structure and operation principle. In this study, a zeolite 13x-assisted planar microwave microstrip resonator sensor, reinforced by regenerative feedback loop, was investigated for CO_2 and CH_4 detection. Real-time monitoring was performed when the sensor was exposed to 1 % to 50% gas concentrations. The sensing was completed by detecting the adsorbents permittivity change during adsorption of the target gases on zeolite 13x. The change in the permittivity was due to the cumulative effect of increased temperature (heat of adsorption) and occupied pores of the adsorbents. By correlating the resonant frequency shift with the concentration of the target gas, the sensor was able to detect CO_2 and CH_4.

471350 Performance of Vibronic Level Switches in Corrosive Services

Abstract:

Vibronic level switches are capable of detecting the presence of liquid and solid products. More generally, the principle of operation of a vibronic level switch provides differentiation of a dense phase from a lighter phase, typically a gas or vapor. Low energy vibration at a fixed frequency is imparted to the metallic level switch probe when it is contacted by the lighter phase, and the vibrational frequency is reduced when a contacting dense phase adds to the mass of the system.

In corrosive services, the metallic probe may lose mass over time due to loss of corrosion by-products and therefore may realize a higher vibration frequency during time in service. The microprocessor for the level switch can include diagnostic alarm settings to indicate an abnormally high vibrational frequency due to corrosion damage.

In highly corrosive services, a corrosion-resistant liner can be applied to the metallic probe. The corrosion-resistant liner, by itself, adds to the mass of the system. Diffusion of corrosive chemicals through the liner material during time in service will result in a slow but finite corrosion rate to the metallic probe. Rather than losing mass over time due the corrosion mechanis, a lined level switch will gain mass over time and will therefore realize a reduced vibration frequency over time. A sufficiently low vibration frequency will be interpreted by the microprocessor as a "tripped" condition; in other words, a condition of being wetted by a dense phase. Lined level switches can be monitored periodically to determine if slow corrosion is impacting the vibration frequency of the instrument.

Processing facilities which handle highly corrosive chemicals must understand the potential for false trips due to corrosion to lined level switches. A false trip due to corrosion to a lined level switch can give the appearance of a unreliable instrument, which can cause operating personnel to doubt the validality of level switch instrument signals.

467127 Gas Phase Composition Measurements Using Ultrasonics

Abstract:

Currently commonly used techniques for gas (vapour) composition measurement are unable to satisfy all the requirements of industrial and laboratorial applications. These techniques include IR, NIR and GC can be both expensive to implement, require calibration and in some cases are not suitable for rapid on-line measurements.Â Speed of sound (SoS) is a fundamental property of a gas (vapour) system and has been demonstrated to be functionally dependent on gas phase composition. This project has deployed a novel speed of sound sensor which can be directly applied in real industrial problems for online composition measurement of binary gas mixtures (e.g. CO_2-N_2) and quasi-binary gas (vapour) mixtures (e.g. dry air-octane). Argon is used to calibrate the cellâ€™s pathlength as its SOS data are well published and not frequency dependent. This presentation will discuss the fundamentals of the instrumentation design, data analysis and the performance of the sensor systems using a series of model gas mixtures.

Considering all of the factors which could cause inaccuracy, the standard uncertainty of this SOS measurement is estimated in this instrument to be < ±0.01 m/s. A temperature-speed of sound-composition database has been constructed for 20 industrially important air-vapour and several binary gas systems over a temperature range from 5 to 55 °C. The SOS across the temperature range of 5 to 55 °C and the gas composition has also been independently determined using a highly accurate gas buoyancy method. The average deviation in accuracy between the predicted and the measured composition for 15-55 °C is typically < 0.5%, and for specific test temperatures can be <0.1%. Typical precision is 1: 300000. In comparison with traditional measurement techniques, this technique is not only robust but also has the potential to be faster, simpler, more cost-effective, more precise and more accurate.

471858 Development of Polymer Sensors for the Online Detection of Waterborne Uranium

Abstract:

Nuclear nonproliferation efforts and treaty verification require portable, robust radiation detectors capable of detecting trace-levels of radionuclides in environmental matrices. A recent development in environmental sensing is a portable, flow cell detector that utilizes extractive scintillating resin. Extractive scintillating resin serves the dual purpose of (1) concentrating the radionuclide of interest and (2) serving as a radiation transducer. Previous resins were produced by physically absorbing organic extractants and fluors in a polymer matrix, which resulted in unstable resins as the active components leach from the polymer over time.

This contribution describes our work to synthesize a new class of extractive scintillating resin in which the fluor and the ligand are bound covalently within the polymer matrix. Two approaches have been taken to synthesize these resins: (1) synthesis of a 4-vinylbenzyl chloride containing polymer followed by solid-phase synthesis techniques to add the uranium-selective ligand and (2) one-pot polymerization utilizing a polymerizable fluor and ligand. In the first approach, suspension polymerization was used to prepare resin beads comprising 4-vinyltoluene; divinylbenzene; the fluor, 2-(1-napthyl)-4-vinyl-5-phenyloxazole; and the solid phase reactant, 4-vinylbenzyl chloride then the resin was functionalized with methylphosphonic acid. In the second approach, resins were prepared by suspension polymerization comprising 4-vinyltoluene; divinylbenzene; 2-(1-napthyl)-4-vinyl-5-phenyloxazole; and the ligand, ethyleneglycol methacrylate phosphate. Porosity of the resin was varied by using different dilutents during suspension polymerization and was analyzed by BET. Resins were characterized by Fourier-transform infrared spectroscopy to support functionalization. Fluorescence properties were studied with spectrofluorometry to measure emission wavelength and intensity. The detection efficiency of the resins in near neutral pH water was evaluated by sequentially loading uranium onto a resin-packed column followed by standard scintillation counting of the column. Resins synthesized by the first approach detected uranium-233 with an efficiency of 40% in pH 4, distilled water. The flow

cell efficiency of the resins was evaluated in an online, flow-cell detector. Resins synthesized by the first approach showed 20% flow cell efficiency in pH 4, simulated ground water. Results from this research are laying the groundwork for the development of portable radiation sensors capable of online monitoring of waterborne alpha-emitting radionuclides.

471343 Emerging Methodologies for Environmental Exposure Assessment: Coupling Personal Sensor Data and Agent Based Modelling (ABM)

Abstract:

Innovations in sensor technology create possibilities to collect environmental data at unprecedented depth and breadth. With the advent of GIS, GPS to track individuals, and personal environmental monitoring, undertaking such analyses throughout an individual's routine, or even lifetime, is now possible. Finding the right balance between limited amounts of high quality data from standardized environmental monitoring campaigns and large amounts of moderate quality data by sensor networks can transform the way we understand and interact with our environment. Measuring, though, personal exposure directly requires a large number of people and therefore is often not feasible due to time and financial constraints. Considering the substantial technical and ethical hurdles involved in collecting real individual space-time movement data for whole populations, this study aims to develop an individual exposure model, where movement and interaction behaviour are simulated using Agent Based Modelling (ABM), informed by personal sensors data. This approach, thoroughly explained below, allows the extrapolation of data from a local scale sensors campaign to the larger population of a region and can produce data that could fill in gaps that exist in traditional datasets.

A trial multi-sensor campaign took place in Thessaloniki, Greece, where 30 participants carried a series of devices such as: (a) a temperature logger, to detect changes between indoor/outdoor conditions, (b) a fitness tracker, to capture motion and intensity of activity, (c) a GPS device, to track coordinates and speed along with (d) Moves, a smartphone application that enables tracking of location and activity. Additionally, a time – location – activity diary was filled out on paper each day. Aiming to substitute cumbersome time logs and to overcome the major drawback of the GPS-type sensors, that fail to differentiate whether an individual stay indoors, outdoors or in transit, the capability of predicting the type of location based ***solely*** on sensor data was explored using an Artificial Neural Network (ANN) model. A multi-layer perceptron network was utilized, which is a feed-forward artificial neural network model that maps sets of input data onto a set of appropriate output. Specifically, the independent variables that fed the ANN input layer consisted of (a) the personal temperature, ***Temp***, derived from the wearable temperature sensor, (b) the observed outdoor temperature, $\boldsymbol{Temp_{out}}$, derived from a central meteorological station of Thessaloniki, (c) the ratio $\boldsymbol{Temp / Temp_{out}}$, and (d) personal speed, derived from the GPS devices wore by the participants. Moreover, information on day light was also included as an input variable, transformed into a categorical element (day or night). The initial database was divided into training and validation set (85% and 15% of the total record entries, respectively) and the models developed from the training set were tested using the validation set. Results illustrate that the ANN model performs very well in predicting the various locations, especially the indoor ones, which also comprise the vast majority of registries of the training set, since most of the time of the daily activity is spent in indoor locations.

As a next step, using Monte Carlo analysis, distributions of participants' movement and activities were extrapolated to a larger population. These data were used as input to a spatially explicit ABM computational platform in order to model trajectories of individual citizens. ABM is a simulation technique where a system -in this case a city- is modelled as a collection of individual heterogeneous actors, the agents. It is a stochastic approach that enables the observation and prediction of emerged behaviours. Every entity taking part in this system is ***agentified***, it is therefore considered as an individual agent (the system is composed of roads, buildings, vehicles and human agents). Specifically, imported road and building shapefiles of a building block resolution contain spatial information on the capacity of a street and land use information (e.g.: type of building) respectively. The extrapolated observations and distributions on spatiotemporal behaviours, derived from the sensors campaign, together with population data from local authorities were transformed into coding lines that define and shape the agents' world. Particular emphasis is also given in the case of in-model incorporation of socio-economic status (SES) data. SES variables (e.g., information on educational level, income, occupational status) can explain differences in external exposure because of different prevalence of specific preferences and decision-making in various groups of population. Different human agents based on different age, sex or income will follow different rules, will express different behaviours and this could lead to a different exposure profile. Moreover, knowledge of human agent characteristics by other human agents provide a signal that acts to enable or prevent interaction from occurring. Data on lifestyle/behaviour patterns (e.g., timetables for various activities per gender and age group) and SES information, derived from censuses in the city of Thessaloniki, were implemented in the human agents' population. Survey outputs were associated with human agent behavioural rules, with the aim to model representative to real world conditions.

When the model is initialized, human agents are clustered in age groups, depending on their age and are randomly allocated to a residential place, which serves as their house for the whole simulation. Human agents' characteristics provide capabilities or constraints on the agents' behavioural rules. The reference point could be, for example, income that can influence their preferences and decision-making. Based on their age and SES characteristics, and of course based on the distance between point of departure and their targeted destination, human agents will choose different means of transportation. In the same way, different human agents will follow a different sequence and types of activities. For example, children and adults are programmed to move from a household to an assigned school or office whereas human agents that belong to the elderly will follow a different sequence of activities. At the end of a run (a single run corresponds to a typical day), human agents' trajectories, derived by the coded routine, are captured as points and exported as a GIS shapefile, a layer which can then be superposed onto high spatial resolution urban air quality maps of hourly PM10 concentration of pollutants, for the city of Thessaloniki. Personal exposure, expressed as inhalation-adjusted exposure to air pollutants is then evaluated by assigning pollutant concentrations to a human agent based on his/her coordinates, activities and the corresponding inhalation rate.

By estimating the daily time-activity patterns (predicted by the coupled sensors-ANN-ABM platform) of vulnerable subgroups of population, we were able to estimate their personal exposure and intake dose per body weight. On average, personal exposure results were between 10 and 20% more accurate than the equivalent estimate using ambient air concentration of PM as exposure proxy.

An individual exposure model was developed, where societal dynamics are being explicitly taken into account. The model can feed into a population-based exposure assessment without imposing prior bias, but rather basing its estimations onto emerging properties of the agents' system behaviour. The establishment of an approach with the capacity for aggregation and analysis at various levels of population size, together with the integration of SES indicators can lead to an exposure assessment model that can be useful especially for vulnerable groups of population, such as children, the elderly and people with low SES. The dynamic nature of intake dose assessment at the individual level allows for the derivation of guidance regarding behavioural options that limit exposure to high levels of pollution. This study represents the first step of the development of a methodological approach towards a refined external exposure assessment and urban air pollution and climate change mitigation management. The model can be further used as a means for estimating and comparing the probable effects of different public health strategies prior to implementation, therefore reducing the time and expense required to identify effective policies.

470233 Heavy Metal Detection in Treated Water Via a Fabric Nanocomposite Sensor

Abstract:

We present the conception of an electrochemical sensor for on-site/portable water analysis. This flexible sensor was developed to quantify the amount of heavy metals (in real-time), alleviating the wait time, large sample size, possible contamination, and expensive analytical equipment associated with current procedures for water analysis. This sensor is composed of a multiwalled carbon nanotube (MWCNT) functionalized polymer mat, for a flexible yet conductive sensor. The MWCNTs are then functionalized with specific molecules which show high selectively towards the target metal ions. The optimization of MWCNTs and selective molecules functionalization, as well as the selectivity and sensitivity of the sensor will be discussed.

451179 Quantitative Analysis of Viral Particles By Tunable Resistive Pulse Sensing Technology

Abstract:

Quantification of viral particles is a crucial issue in both virology research and biotechnology applications. However, characterization of viral particles remains challenging because of their nanoscale dimensions, as well as some limitations of conventional technologies. In recent years, tunable resistive pulse sensing technology has been developed rapidly as an emerging and useful tool for detection of individual nanoparticles. This study reports particle size distribution, concentration and zeta potential characterization of various viral particles, including tobacco mosaic virus, baculovirus, bacteriophage, and so forth. The results showed that the data of virusâ€™ size was comparable to that from other measurement methods. The total viral particle concentrations were also obtained, which allowed comparison with previous infectious titer assay results. The zeta potential measurements provided additional surface charge information of viruses that has never been reported. This work enabled us to have an in-depth understanding of individual viral particles.

463109 Multifunctional Fluorescent Optical Sensors for Bioanalysis

Abstract:

Optical sensors are important tools for cellular imaging and analysis, for probing cellular metabolism, for understanding pathways and biological and physiological processes, and for diagnosing diseases and cancers. We have developed several series of fluorescent optical sensors (pH, oxygen (O_2), potassium ion (K^+), and glucose sensors) suitable for intracellular and/or extracellular imaging and analysis. We integrated individual sensors as dual sensors using polymeric synthesis techniques for simultaneous multi-parameter measurements. Of particular value for application in the complex biological environment, some of the sensors emit two or even three colors in response to the same excitation wavelength, enabling ratiometric measurements and providing superior measurement accuracy. For example, we have prepared a dual pH/O_2 sensor with three emission colors. The tri-color sensor composed of a blue emitter as an internal build-in reference probe, a green emitter as the pH probe, and a red emitter as the oxygen sensor. The build-in reference probe is unresponsive to pH or O_2. It was demonstrated that this polymer-based tri-color sensor is capable for measuring the changes of pH and oxygen concentrations in photosynthesis cyanobacteria possessing high autofluorescence. We also integrated glucose and oxygen sensor as ratiometric dual glucose/oxygen sensors for analysis the glucose and oxygen consumption for bacterial cells and mammalian cells in real time. This dual glucose/oxygen sensor enables simultaneous measurement of glucose and oxygen, providing important information for continuous monitoring of glucose, diagnosing diabetes and hypoxia (low oxygen) related diseases and cancer, as well as the understating of biological processes of the metabolism.

467999 *An Ammonia Sensor for a Handheld Analyzer for Non-Invasive, Real-Time Monitoring of Kidney and Liver Disorders*

Abstract:

A small device that could reliably measure urine ammonia in either the outpatient or inpatient settings would have a myriad of clinical applications. Diagnosis and management of disorders that impact absolute or relative changes in renal excretion of ammonia, including conditions that alter systemic acid/base homeostasis and diseases of the kidney and liver, would be greatly aided by such a device. However, traditional practical and technical limitations of urine ammonia measurement have impeded the use of urine ammonia as a routinely employed biomarker. This project is focused on the development of analyzer technology directed at ammonia measurement and the implementation of this analyzer into a practical handheld device prototype. The proposed technology is based on a unique hybrid sensor array capable of quick, precise, and reliable measurement. The analyzer will be capable of being applied in two broad formats: 1) a handheld device for outpatient or inpatient applications requiring less frequent (e.g., daily) monitoring, and 2) a urinary catheter device for semi-continuous measurement for inpatient applications requiring intensive serial monitoring. The miniaturized, low cost, easy to use handheld device prototype will utilize solid-state electronics with a high level of integration. Furthermore, the prototype will connect wirelessly to a centralized data display and storage unit (e.g., cellular phone), allowing a clinician to easily track an individual or a group of patients. This work shows the prototype of the sensor, and its capability for ammonia quantification in biological samples in real-time scale. The sensor shows fast and reversible response.

467570 Rapid, Continuous Monitoring of Protein Levels in Food Matrices By Combining Microfluidics and Printed Electronics

Abstract:

Ensuring the safety and quality of food products requires a rapid and accurate diagnostic platform that operates with minimal user input when measuring toxin or pathogen levels. We have developed such a biosensor by combining printed electronic and microfluidic technologies to measure the concentration of ricin, a toxic protein, in aqueous buffer, orange juice, and milk. The label-free detection strategy transduces protein binding at a floating-gate electrode via a previously presented mechanism [1] giving a quantitative electronic readout that correlates with the amount of protein in solution. Specificity for ricin is achieved by functionalizing the gold floating gate with an aptamer that is selective to the target protein ricin but the underlying mechanism is interchangeable with other protein or nucleic acid targets. The principal advantage is the ability to decouple the electronic transducer from the microfluidic capture interface so that the protein can be measured from complex matrices such as orange juice and milk. This feature allows the sensor to be configured in-line with current manufacturing processes thus providing a continuous monitor for toxin levels, an example of which is ricin.

We were able to achieve limits of detection as a low as 1 ng/mL in buffer and 10 ng/mL in milk and orange juice by operating the printed electronic device at 13 Hz, rather than the common DC limit. This AC operation improves the signal-to-noise allowing for smaller amounts of bound protein to be identified. Additionally, the parallel fabrication strategy and microfluidic control of complex fluids minimizes the background noise from device-to-device variations, electrode instability, and non-specific adsorption. As little as 1 ng of ricin is necessary to generate a signal that arises in less than 45 minutes. These performance metrics demonstrate the potential for this technology to impact the food safety industry by providing manufacturers with more information from less time/labor leading to safer products sent to the public.

465959 Aptameric Peptide Mediated Capacitive Detection of Protein Kinase

Abstract:

Abstract- This work proposes a flexible microchip with electrolyte-insulator-semiconductor (EIS) based detection of protein kinase using an aptameric peptide. In the presented device, the protein kinase A specific aptamers are linked to gold nanoparticles, anchored on the SiO_2 surface. On addition of protein kinase A, the aptameric peptide binds to it. The approach is based on the modulation of charge upon aptamer-kinase interaction on the pH and charge sensitive oxide-semiconductor surface. The enzyme concentration was evaluated by measuring the resulting change in the flat-band voltage of the EIS sensor.

Introduction- The phosphorylation of proteins regulates almost all aspects of cell life. Protein kinases are enzymes which catalyze the phosphorylation of target proteins. Unusual regulation of protein phosphorylation is associated with many diseases, including cancer. Therefore, accurate quantification of the protein kinases is crucial for the analysis of related abnormalities.

Recently, kinase activity assays based on mass spectroscopy, fluorescence microscopy, and quartz crystal microbalance have been reported for the analysis. However, miniaturization of these techniques for handheld device is quite not possible. Î¼TAS technologies have been widely explored to develop miniaturized devices for portable use. Recent advances in microfabrication technique have facilitated the creation of microchips with electrochemical detection of different biomarkers. To detect protein kinases electrochemically, a variety of methods have been proposed. Major drawback in these reports is the use of multipleÂ bio and nano particles for sensitive and selective detection. These steps make the analysis laborious and expensive.

For this reason, this work focuses on the development of a microchip for protein kinase A (PKA) detection with a minimalistic use of biomolecules. Electrolyte-insulator-semiconductor (EIS) devices as electronic transducers for label-free detection of biochemical reactions are very promising for the development of low-cost high-throughput sensors. The capacitive EIS device is the simplest field-effect bio-chemical sensor with a direct electrical readout. The change in the local pH at the electrolyte-insulator interface or the attachment of charged biomolecules to the insulator alters the surface potential, leading to a shift in flat-band voltage (V_{fb}).

Therefore, this work, uses an EIS capacitor structure to assay the protein kinase activity. A peptide based aptamer was immobilized on the insulator surface using a (3-Mercaptopropyl)trimethoxysilane (MPTES) and gold nanoparticle (AuNp) link. The EIS sensor detects the charge changes in AuNp/aptamer induced by the binding of PKA and aptamer. The concentration of enzyme is detected in terms of changes in the flat-band voltage of the EIS sensor.

Experimental- 50 nm of SiO_2 was deposited by chemical vapor deposition (CVD) on p-type Si wafers. The wafers were then cut into small pieces of 8 x 8 mm in size. The Si-SiO_2 wafers were cleaned using ultra-sonication in acetone and isopropyl alcohol for 20 min, each. After drying the wafers, 100 nm aluminium was deposited on the back of the Si wafers to serve as an ohmic back-contact, using a vacuum thermal evaporator.

The cleaned sensor surface was first silanized by dipping it in a freshly prepared 1% MPTES in ethanol for 24 hrs. The surface was then thoroughly cleaned using ethanol. For deposition of AuNPs on the sensor surface, the thiol functionalized surface was made to react with an AuNPs colloid for 12 hrs.

A rapid prototyping technique for microchip based on a polymer film and double sided tape was developed. A lower layer consists of an EIS sensor mounted on a PES film and attached to an Au electrode. The middle layer was fabricated from 3M^{TM} double sided tape and contained the fluidic network and reservoir for the analysis. Finally, the upper layer consists of an Ag/AgCl thin-film reference electrode patterned on the PES film.

Results- Each modification step of the sensor surface was characterized by means of water contact-angle measurements with a 12 Î¼l drop of ultrapure water. The hydrophilicity of the surface increased significantly to a contact angle of ~7.8Â° after the UV-O_3 treatment from a contact angle of ~42Â° for a bare SiO_2. The increase of surface hydrophilicity is due to the increase in hydroxyl groups. The sensor surface modified with MPTES regained the hydrophobicity (contact angle= ~37Â°). The change in the surface property of SiO_2 confirms the functionalization of the sensor. A scanning electron microscopic analysis of the EIS surface was also performed to check the presence of AuNPs.

Binding of PKA enzyme with aptamer altered the surface charge, leading to the shift in V_{fb}. A linear shift in V_{fb} was obtained with the increasing concentration of PKA enzyme. Based on these results, we propose a flexible microchip with an EIS sensor for the label-free detection of PKA enzyme.

Conclusion- EIS based detection of PKA using a peptide aptamer is presented in this work. A label-free analysis without the use of any reporting molecule was performed. The proposed device can easily be modified to detect any other PK using specific aptamers. Polymer based chip and one-step detection makes this work more practical. This assay offers sensitive and reproducible quantification of PK.

467401 Highly Sensitive Gluten Sensing Platform Based on Conducting Polymers Bio-Functionalized with R5 and G12 Gluten Antibodies

Abstract:

It is estimated that gluten intolerance and celiac disease affect around 1% of human population. Considering the scale of this issue and the lack of treatment, patients with gluten triggered immunologic disorders have to rely strictly on gluten free diet. Depending on the level of advancement inflammatory response in patients can be triggered even at very low (20 ppm and less) gluten levels. The lack of sufficient tests and gluten measuring methods makes it hard for patients affected by this disorder to function. Here we present two devices that determine amount of gluten in food samples. Chemiresitive device based on conducting polymer thin film functionalized with two gluten antibodies R5 and G12 is capable of sensing gliadin; one of the main proteins in gluten; with precision up to 20 ppm. The second device builds on the first one by adding a gate electrode to existing chemiresistor channel. This organic electrochemical transistor not only magnifies the signal, making measurement more sensitive and more precise, but also allows for miniaturization, which in result have great advantage in real applications. Low power consumption and sub-volt operation mode make both devices a great candidates for portable gluten analyzers.

473079 Development of a Mobile CO_2 Breath Sensor for Pulmonary Assessment

Abstract:

Obstructive pulmonary disease is an extremely common disease in the United States, yet lacks heavily in consumer-level medical devices. The measurement of Carbon Dioxide and volumes in breath give direct indicators of pulmonary obstruction, but require very rapid and non-intrusive analysis, which brings high-cost and user-unfriendly designs. We have developed an inexpensive and user friendly sensor which takes advantage of colorimetric carbon dioxide detection, the Confined Pitot Tube flow meter, and precision temperature compensation in a smartphone-based device. With the use of advanced regression methods, we have made a low-cost device that is capable of accurate analysis of characteristic parameters, such as End-Tidal carbon dioxide and volume, along discriminating minute details in carbon dioxide breathing waveforms. The colorimetric detection platform utilizes a highly porous substrate and ultrafast carbamate reaction system in order to provide CO_2 response times on the order of less than 100ms, with simple and robust transmission LED/photodiode electronics for signal acquisition. The Confined Pitot Tube flow meter allows for accurate flow measurements at a low back-pressure for optimal breath flow measurements using an inexpensive differential pressure sensor. Using a precision temperature model, we are able to compensate for large temperature swings which otherwise bring large deviations in sensor readings. Combining these technologies, we are able to demonstrate an inexpensive, robust, and user-friendly device that has a demonstrated capability to discriminate between varying degrees of lung obstruction, providing a new tool for monitoring chronic obstructive disease and a potentially very powerful tool for monitoring episodic pulmonary disorders, such as Asthma.

462692 A Wearable Electrochemical Platform for Non-Invasive Simultaneous Monitoring of Ca^{2+} and pH

Abstract:

Wearable biosensors have become important for ambulatory care to promote individuals' role on their health by employing personal health system. Recently, many research groups have developed wearable devices for real-time monitoring of various electrolytes and metabolites non-invasively using tattoo and textile-based sensors. Despite recent development of epidermal analysis in body fluids, real-time Ca^{2+} and pH detections have not been fully explored. Homeostasis of ionized calcium in biofluids is critical for human biological functions and organ systems. Measurement of ionized calcium for clinical applications is not easily accessible due to its strict procedures and dependence of pH. pH balance in body fluids greatly affects metabolic reactions and biological transport systems. Here, we proposed a wearable electrochemical device for continuous monitoring of ionized calcium and pH of body fluids using a disposable and flexible array of Ca^{2+} and pH sensors integrated with a flexible printed circuit board (FPCB). This enables real-time quantitative analysis of these sensing elements in body fluids. Accuracy of Ca^{2+} concentration and pH in biofluids measured by the wearable device is demonstrated. Our results show that the wearable sensors are highly sensitive and selective to target ions with high repeatability and stability. Real-time on-body assessment of sweat is also performed, and results indicate that calcium concentration increases with decreasing pH. This platform can be used in non-invasive continuous analysis of ionized calcium and pH in body fluids for disease diagnosis such as primary hyperparathyroidism and kidney-stone diseases.

471326 Electrochemical Sensors for the Rapid Detection of *Pseudomonas Aeruginosa* in Polymicrobial Environments

Abstract:

A major cause of death among hospitalized patients results from infections acquired within the hospital setting. Over 1.4 million people experience infection complications each year, most notably from surgical wounds and pneumonia. *Pseudomonas aeruginosa* is a bacterial pathogen commonly found in these nosocomial infections. *P. aeruginosa* produces a small tricyclic molecule known as pyocyanin and because it is redox-active, it can be detected using electrochemistry. As a unique by-product of *P. aeruginosa*, pyocyanin is a useful biomarker for identification and detection of this opportunistic pathogen. It has been previously shown that electrochemical sensors can be used to detect *P. aeruginosa*â€™s production of pyocyanin in complex media environments. However, further research is needed to show how *P. aeruginosa* behaves when co-cultured in a complex environment with other bacterial pathogens. This study addresses this need by electrochemically monitoring *P. aeruginosa*â€™s production of pyocyanin in polymicrobial samples.

A total of 5 liquid cultures of different bacteria (*Pseudomonas aeruginosa, Escherichia coli, Staphylococcus aureus, Staphylococcus epidermidis*, and *Enterococcus faecalis*) were grown overnight at 37 Â°C in lysogeny broth (LB) or trypticase soy broth (TSB) growth media. From these stock cultures, different polymicrobial combinations were tested, starting with single species inoculated with *P. aeruginosa*. From there, combinations of three bacterial pathogens were tested including a final sample that contained all five bacterial species. Production of pyocyanin was monitored every hour for the first 24 hours and then every 5 hours afterwards for a total of 3 days. Electrochemical detection was completed with 100 ÂµL of liquid sample placed onto a disposable, screen-printed sensor. Measurements were recorded using square wave voltammetry at an amplitude of 0.05 V and a frequency of 15 Hz.

Measurements of current versus applied voltage were recorded for each of the different bacterial samples. Using a calibration curve of pyocyanin standards, the measured current can be correlated to the pyocyanin concentration in the sample. The study found that *P. aeruginosa* produces pyocyanin at similar rates, regardless if other bacterial pathogens are present. This data is useful towards developing a rapid point-of-care diagnostic sensor for *P. aeruginosa* in healthcare-associated infections that contain a number of different bacterial pathogens.

464852 Bioactive Block Co-Polymer Interfaces for High-Avidity Pathogen Capture and Enrichment

Abstract:

Detection of microbial pathogens from environmental or clinical samples often requires pathogen isolation and enrichment from complex media. Using lectins as capture agents for this application is advantageous, as lectin-functional surfaces directly isolate microbial cells through specific binding interactions with exopolysaccharides (EPS) moieties, can be used to catch and release targeted cells, and are inexpensive relative to capture antibodies. However, lectin-based capture strategies are generally limited by inherently weak oligosaccharide binding affinities, ultimately resulting in poor capture efficiency. In this work, we have developed a block co-polymer for clustering high concentrations of lectin capture proteins together on interfaces to promote multi-valent binding interactions that drive high-avidity microbe capture. The block co-polymer poly(glycidyl methacrylate)-*block*-poly(vinyl dimethyl azlactone) (PGMA-*b*-PVDMA) was synthesized to controlled block lengths using a RAFT polymerization. This polymer is chemoselective, with epoxy groups present in GMA block that are reactive towards hydroxyl groups present at interfaces, and azlactone groups present in VDMA blocks reactive towards capture proteins. Upon surface functionalization, the length of the PVDMA block modulates the chemical reactivity of the surface, and the high densities of reactive azlactone groups resulted in the loading of capture lectins at up to 50-fold higher density compared to alternative coupling chemistries. These lectin-functional polymer interfaces were able to capture up to 43% more microbes from solution compared to control surfaces. By applying techniques in micro and nanofabrication, these polymers were patterned into microscale surface structures on silicon interfaces, allowing for identification of structures that favored the capture of single cells or of cellular aggregates of controlled size. Further optimization of these interfaces will allow for a combination of optimized microscale architecture, nanoscale roughness, and lectin density to achieve optimal bacterial pathogen isolation from food, water, or clinical samples.

462701 A Wearable Microsensor Array for Multiplexed Heavy Metal Monitoring of Body Fluids

Abstract:

A flexible and wearable microsensor array is described for simultaneous multiplexed monitoring of heavy metals in human body fluids. Zn, Cd, Pb, Cu, and Hg ions are chosen as target analytes for detection *via* electrochemical square wave anodic stripping voltammetry (SWASV) on Au and Bi microelectrodes. The oxidation peaks of these metals are calibrated and compensated by incorporating a skin temperature sensor. High selectivity, repeatability, and flexibility of the sensor arrays are presented. Human sweat and urine samples are collected for heavy metal analysis, and measured results from the microsensors are validated through inductively coupled plasma mass spectrometry (ICP-MS). Real-time on-body evaluation of heavy metal (*e.g.* zinc and copper) levels in sweat of human subjects by cycling is performed to examine the change in concentrations with time. This platform is anticipated to provide insightful information about an individualâ€™s health state such as heavy metal exposure and aid the related clinical investigations.

471325 Electrochemical Detection of *Pseudomonas Aeruginosa* in Human Wound Exudate for Point-of-Care Applications

Abstract:

A significant limitation to antibiotic stewardship and improved patient care is the delay between the obtainment of a biological sample and its bacterial identification in the clinical setting. The use of plate cultures inoculated from swab samples is currently the standard practice for positive identification. Because this process can take anywhere from 24 to 48 hours before a positive result can be produced, there is an unmet need to develop rapid alternatives for bacterial identification. In clinical care, *Pseudomonas aeruginosa* is one of the leading causes of bacterial infections among patients with cystic fibrosis or compromised immune systems. Given the increasing prevalence of this highly resistant pathogen and the significant delay in care as a result of current bacterial identification practices, a rapid alternative would allow a physician to switch from broad-spectrum antibiotics to more direct targeted therapy, lowering antibiotic resistance and improving patient care outcomes.

To address this need, we report the use of an inexpensive, electrochemical sensor to detect pyocyanin, a unique, quorum sensing molecule secreted by *Pseudomonas aeruginosa*, in human wound exudate obtained from patients with chronic wounds. Because pyocyanin is redox-active, it can be detected using electrochemical sensors. This method eliminates sample preparation, takes less than two minutes to perform, and only requires 7.5 microliters of sample to analyze.

A total of 14 wound fluid samples were obtained from 12 patients enrolled in the Wound Etiology and Healing Study (WE-HEAL) at George Washington University. Wound effluent specimens were collected from swabs using the Levine technique. After collection, the swabs were placed in 0.65 μm pore size filters and centrifuged at 12,000 rpm for 4 minutes to remove cellular and fibrinous debris and to extract wound exudate. The samples were then stored at -80 °C until analysis. For each test, 7.5 μL of wound exudate was pipetted onto a screen-printed electrochemical sensor

and square-wave voltammetry was used to determine the presence or absence of pyocyanin in the samples. The electrochemical results were compared against 16S rRNA profiling, one of the gold standards for bacterial identification. The electrochemical results yielded 9 correct matches, 2 false negatives, and 3 false positives, giving a sensitivity of 71% and specificity of 57% for detection of *Pseudomonas aeruginosa*. The results from this study will be used in improving the current sensor design and validating the use of an electrochemical sensor as a point-of-care diagnostic for detecting *Pseudomonas aeruginosa*. More importantly, the sensor can be functionalized to potentially detect other bacterial species by altering the applied potentials and thus can be enhanced into a point-of-care device with the capability to detect a wide range of clinically relevant bacterial pathogens.

460964 Programmable Living Material Containing Reporter Micro-Organisms Permits Quantitative Detection of Disease-Relevant Molecules

Abstract:

The increasing molecular understanding of many diseases permits the development of new diagnostic methods. However, few easy-to-handle and inexpensive tools exist for common diseases such as food disorders. Recently, bioinspired stimuli-responsive materials containing living organisms have been reported [1, 2]. By enclosing micro-organisms into a sandwich consisting of a solid bottom and a size-selective porous top polymer sheet, the organisms could not escape but were still able to interact with the outer environment. Based on this, we developed a whole-cell biosensor platform that combines the diffusion behavior of an analyte with an analyte-specific bacterial fluorescence reporter system [3]. The analysis requires only a photo camera, a blue-light source and a blue-light optical filter sheet. The well-established model bacterium ***Escherichia coli,*** genetically modified with a fluorescent reporter system, was incorporated in a matrix of different polymers to create a programmable living material-based analytical sensor (LiMBAS). This material shows an easily quantifiable and specific response when coming into contact with dissolved analyte anywhere on its surface. In a realistic application, we programmed LiMBAS to detect lactose directly from cow's milk and galactose from lactose-free milk without previous dilution. We chose milk for our proof-of-principle experiment since it is a widely used food product and a source of several common food disorders including lactose intolerance, allergies as well as galactosaemia. LiMBAS could accurately quantify lactose or galactose in undiluted food samples and was able to measure food intolerance relevant concentrations in the range of 1–1000 mM requiring a sample volume of 1–10 μL. It worked at ambient temperature and humidity conditions and was storable ready-to-use for more than one week. We additionally discuss how the quantification principle of combining diffusion behavior with matrix-embedded microbial whole cell reporter systems could be applied to develop a variety of biomedical diagnosis tools for the analysis of food, blood or environmental constituents that can be used in domestic or outdoor applications. A wide range of genetic tools for ***E. coli*** are readily available thus allowing the reprogramming of the material to serve as biosensor for other molecules. LiMBAS as biomaterial offers the possibility to safely use engineered micro-organisms outside of a laboratory environment without the need for culturing or complicated analysis. In combination with smartphones, an automated diagnostic analysis becomes feasible which would also allow untrained people to use LiMBAS.

[1] L.C. Gerber, F.M. Koehler, R.N. Grass, W.J. Stark, Proc. Natl. Acad. Sci. U. S. A 109 (2012), pp. 90–94

[2] L.C. Gerber, F.M. Koehler, R.N. Grass, W.J. Stark, Angew. Chem. Int. Ed., 51 (2012), pp. 11293–11296

[3] C.A. Mora, A.F. Herzog, R.A. Raso, W.J. Stark, Biomaterials, 61 (2015), pp. 1-9

463196 Introduction of Two Electrical Impedance Systems for the *in-Vitro* Characterization of Huvecs Undergoing Hydrodynamic Shear Stress

Abstract:

Kinetics of endothelial cells undergoing hydrodynamic shear stress are extremely dynamic and require a real-time evaluation. Electrical impedance systems have shown to have the ability to indicate numerous status factors of cellular monolayers in real-time, including cell viability [1], confluency [2], cell-cell integrity and morphology [3]–[5]. Within this work, we introduce two electrical impedance chips for the in-vitro characterization of HUVECs undergoing hydrodynamic shear stress. Each system has a different fluid dynamic and structure environment that can cater to the scientist investigation interest and cell inoculation preference.

One system is a microfluidic platform where steady laminar shear stress is generated within a vessel-like structure. The microfluidic impedance platform consists of two microfabricated arrays of independently addressable gold electrode of various diameters (50 µm, 100 µm, and 200 µm), and a common large counter electrode on a borosilicate substrate. Different electrodes sizes were used to test and compare the behavior of different cell population sizes. The microfluidic chip also includes a two-channel design, such that one may contain cells, while the other remains without cells. This design feature helps to identify environmental variables affecting the endothelial cell data.

The other system, an orbital shear platform consists of a 35 mm diameter open-well formed by a PDMS mold. This design allows for more simplified and conventional cell inoculation methods. Unlike the steady laminar shear stress observed in the microfluidic platform, within the orbital shear platform, oscillatory hydrodynamic shear is reproduced. Oscillatory shear is more representative of the pulsatile nature of blood flow and also produces various shear values across the well dish, enabling simultaneous testing of different shear stress magnitudes on the same endothelial cell culture sample. Shear stress was induced by placing the platform on an orbital shaker at a rotation speed of 120 rpm. A computational model was used to determine the temporal and spatial resolution of wall shear stress for one orbit cycle [6]–[8]. Similar to the microfluidic platform, the orbital shear platform consists of microfabricated arrays of 200 µm (in diameter) gold sensing electrodes and complimentary counter electrodes on a borosilicate substrate. Only 200 µm diameter electrode were incorporated into this design due to the complex fluid dynamics which benefit from more measurements of the same subpopulations size for reproducibility purposes. The chip contains 3 rows of electrodes which are spaced 120 degrees from each other. Electrodes are positioned at defined radial locations of 0, 2.5, 5.0, 7.5, 10.0, and 12.5 mm with respect to the center of the borosilicate chip. Within the rotating orbital well, cells experience a range of shear stress values and in differing directions at selected radial positions [6]. Depending on the electrode position, cells experienced shear values between 0.6-6.71 dyne/cm^2.

Both systems have the ability to monitor cell adaptation to hydrodynamic shear stress in real-time via impedance spectroscopy measurements. Chips were inserted into commercially available card readers and connected to an in-house designed printed circuit board programmed to automatically record impedance spectra from a high-precision impedance analyzer (HP4294a) at specified time intervals. Human umbilical vein endothelial cells (HUVECs) were used as an initial endothelium model and cultured within the chips. The collected impedance spectra were fitted to an equivalent circuit model. The individual electrical elements are extracted and used as representative variables for the cellular monolayer and microenvironment behavior. In both systems despite the different fluid dynamic environments, results indicate that the onset of shear stress (acute flow) generates an increase in the cell monolayer resistance, which is followed by a decrease. This is all accompanied by the re-orientation and elongation of cells.

References:

[1] X. Zhang, F. Li, A. N. Nordin, J. Tarbell, and I. Voiculescu, "Toxicity studies using mammalian cells and impedance spectroscopy method," *Sens. Bio-Sensing Res.*, vol. 3, pp. 112–121, 2015.

[2] S. Michaelis, R. Robelek, and J. Wegener, "Studying Cell-Surface Interactions In Vitro: A Survey of Experimental Approaches and Techniques," in *Tissue engineering III: cell - surface interactions for tissue culture techniques.*, vol. 126, 2012, pp. 33–66.

[3] J. Seebach, P. Dieterich, and F. Luo, "Endothelial barrier function under laminar fluid shear stress," *Lab. ...*, vol. 80, no. 12, pp. 1819–31, Dec. 2000.

[4] J. Seebach, G. Donnert, R. Kronstein, S. Werth, B. Wojciak-Stothard, D. Falzarano, C. Mrowietz, S. W. Hell, and H. J. Schnittler, "Regulation of endothelial barrier function during flow-induced conversion to an arterial phenotype," *Cardiovasc. Res.*, vol. 75, pp. 596–607, 2007.

[5] S. Arndt, J. Seebach, K. Psathaki, H. J. Galla, and J. Wegener, "Bioelectrical impedance assay to monitor changes in cell shape during apoptosis," *Biosens. Bioelectron.*, vol. 19, pp. 583–594, 2004.

[6] J. M. D. Thomas, A. Chakraborty, M. K. Sharp, and R. E. Berson, "Spatial and temporal resolution of shear in an orbiting petri dish," *Biotechnol. Prog.*, vol. 27, pp. 460–465, 2011.

[7] A. Chakraborty, S. Chakraborty, V. R. Jala, B. Haribabu, M. K. Sharp, and R. E. Berson, "Effects of biaxial oscillatory shear stress on endothelial cell proliferation and morphology," *Biotechnol. Bioeng.*, vol. 109, no. 3, pp. 695–707, 2012.

[8] R. E. Berson, M. R. Purcell, and M. K. Sharp, "Computationally Determined Shear on Cells Grown in Orbiting Culture Dishes," in *ADVANCES IN EXPERIMENTAL MEDICINE AND BIOLOGY: OXYGEN TRANSPORT TO TISSUE XXIX*, 2008, pp. 189–198.

458208 Engineering a Robust DNA Split Proximity Circuit for Probing Proximal Biological Recognition Events

Abstract:

DNA circuit is a modular and highly-programmable toolbox which can be designed for the autonomous probing of dynamic events, e.g. biomolecular interactions. In this talk, we will introduce a "plug-and-play" DNA split proximity circuit which was engineered to spatially probe for recognition events occurring within close proximity. The central computing module, driven by the concept of association toehold, served as a signal transducer to convert diverse recognition events (e.g. DNA hybridization, aptamer-thrombin binding, small molecule-protein interaction) into readout signals (e.g. fluorophore-quencher and hybridization chain reaction).

We recognize that a major common problem hindering the successful implementation of several in silico circuit designs is the presence of undesired circuit leakage. To ensure the robustness and practical relevance of our proposed design, we focused on identifying the sources and characterizing various types of leakage in order to devise effective counter strategies. Three key findings which are generally applicable to other DNA circuits will be shared in this presentation.

First, we provided direct experimental evidence that 3'-truncated oligonucleotides as the major defected product in solid-state synthesis was the main culprit for initial leakage. Next, we introduced a unique strategy, termed "inter-domain bridging", by translocating a single nucleotide across split domains to eliminate toehold-independent hybridization while improving the strand displacement kinetics across the three-way junction by ca. 13-fold and enhancing the equilibrium signal by one-fold. Finally, we found that asymptotic leakage was inevitable given the inherent dynamics of the intermediate complexes involved in a three-way junction circuit. The best counter-strategy was to use the shortest possible association length, which we recommend to be less than 6 nt.

These three key design guidelines culminated in a final optimized circuit design which was successfully implemented on a model streptavidin-biotin system. The circuit operated robustly against both circuit leakages and interferences from non-specific proteins or a biological environment of 10% fetal bovine serum. A limit of detection of 2.95 nM was achieved within 10 min of detection. We anticipate that this simple signal transduction strategy can be used to probe for diverse biomolecular interactions when used in conjunction with specific target recognition moieties. Successful systems tested in our lab include thrombin-aptamer binding, HER2 homodimerization and DNA hybridization.

459672 Pharmacokinetic Model of a Tissue Implantable Cortisol Sensor

Abstract:

Cortisol is an important glucocorticoid hormone whose biochemistry influences numerous physiological and pathological processes. Moreover, it is a biomarker of interest for a number of conditions, including post-traumatic stress disorder (PTSD), Cushingâ€™s syndrome, Addisonâ€™s disease, and others. Due to both the diurnal cycle and the pulsatile release of cortisol, an implantable biosensor capable of real time monitoring of cortisol concentration in adipose tissue can potentially revolutionize the diagnosis and treatment of these disorders, as well as provide an invaluable research tool. Towards this end, we develop a mathematical model, informed by the physiological literature, to predict dynamic cortisol concentrations in adipose, muscle, and brain tissues, where a significant number of important processes with cortisol occur. The mathematical model was applied to both a healthy adult male, as well as a model Cushingâ€™s disease patient. Values predicted by the model were verified against reported measurements from the physiological literature. Our model can be used to inform the design of an implantable sensor, by optimizing the sensor dissociation constant, apparent delay time, and magnitude of the sensor output versus system dynamics. Measurements from such a sensor would help to determine systemic cortisol levels, providing much needed insight for proper medical treatment for various cortisol-related conditions.

457465 An "Artificial Nose" for the Non-Invasive Diagnosis of Anxiety in Alveolar Breath

Abstract:

An "Artificial Nose" for the Non-invasive Diagnosis of Anxiety in Alveolar Breath

Jessica Fitzgerald and Hicham Fenniri

313 Snell Engineering Center, Northeastern University, 360 Huntington Avenue, Boston, MA 02115

In recent decades, artificial olfactory devices have been developed for disease diagnosis by evaluation of exhaled volatile organic compounds (VOCs) produced from multiple metabolic processes. These devices consist of a cross-reactive sensor array capable of interacting with multiple vapor analytes, a signal transduction mechanism, and response pattern recognition software. Our working hypothesis is that an increase in production of specific VOCs reflecting compromised metabolic processes found in patients suffering with a range of clinically significant anxiety symptoms could be detected with an e-nose. Such a fingerprint approach may also unveil possible biological pathways relevant to anxiety psychopathology.

This paper will present the design and fabrication of our barcoded resin-based (BCR) sensor array and its application for the detection and identification of VOCs. BCRs were prepared from a library of alkylated and fluorinated styrene monomers combined in a binary fashion. Upon interaction with an analyte, the vibrational signatures of the polymer array change, resulting in slight but detectable spectral variations for each BCR. The collective response (or analyte-specific patterns) will then be quantified using multivariate data analysis. This platform improves upon existing technologies as it dramatically increases sensitivity and information content using vibrational spectroscopy of a large library of sensory elements (encoded polymers)

Our current goal is to optimize a sensor array of BCRs for detecting clinically significant anxiety and stress VOCs with highest disease specificity in exhaled breath.

464320 High-Transconductance Ionic Transistors for Ionic Conductance Biosensing Platforms

Abstract:

Non-linear micro/nanofluidic circuit components like ionic diodes and ionic transistors allow precise measurement and control of ion and biomolecule transport in integrated micro/nano total analytical systems which promotes various medical diagnostic applications and single cell measurement technologies. However, most current ionic circuit devices can neither deliver a high ion flux, nor amplify the ion current. High-flux ionic circuits are hence needed to meet the high-throughput requirement of massively multiplexed smart sensor arrays, and autonomous *in vivo* or *in vitro* cell screening platforms. To realize these high-flux ionic circuits, we have previously designed an ionic memristor based on a reversible redox reaction on a silicon electrode [Sun et al, Small, 11, 5206-5213 (2015)]. Here, we report the design of an ionic circuit amplifier and transistor, which allows a small normal gating voltage signal to control a large draining ionic cross current [Sun et al, LabChip,16, 1171-1177 (2016)].

The gating action is based on non-equilibrium ion transport through an ion-selective nanoporous membrane or nanochannel. Because of its charge selectivity, asymmetric concentration polarization occurs across such a membrane upon the application of a normal electric field. The cross current is controlled by the length of the low-conductance depletion front, which must have a strong correlation to the gating voltage to achieve high transconductance. Earlier efforts with nanoslot-microchannel designs have produced a gating-voltage independent depletion front length, either a very short one with a length-scale comparable to the nanoslot dimension [Yossifon et al, Phys Rev E, 81, 046301 (2010)] or a long one that extends all the way to the electrode [Chang et al, Annu Rev Fluid Mech, 44, 401-426 (2012)]. The former occurs when the nanoslot width is small compared to the external microfluidic channel width and the latter when the opposite is true.

To achieve a gating voltage-sensitive depletion front length, we configure the voltages of the draining current such that

the coion flux balance for this draining current, necessary to maintain electro-neutrality, is only possible for a particular depletion length. The depletion front produces an unfavorable concentration gradient for this coion flux at the gating membrane, corresponding to a funnel-shaped energy surface. The size of the funnel and its depth, which control the local coion draining current, is sensitive to the gating voltage. The depletion front length must hence correspond to a specific value to ensure coion flux balance throughout the cross channel. We are hence able to produce a gating-voltage sensitive depletion front length which is insensitive to the external geometries.

This positive feedback mechanism due to a two-dimensional depletion front evolution controlled by two competing electrodes is confirmed by fluorescence imaging and numerical simulation. The membrane ionic transistor chip we have designed based on this principle is able to achieve a transconductance of more than 12 μS, which is the larger than previous reported for ionic transistors by a factor of 40. It corresponds to a voltage amplifier with an amplification factor of 10. The ion current amplification design will be shown to improve a particular nucleic acid ionic conductance membrane sensor by orders of magnitude with fM sensitivity and a tunable 3 orders dynamic range that can be shifted with the adjustable limit of detection. More details of the bipolar membrane sensor design will be presented in another talk in Session T3003.

AUTHOR INDEX

AUTHOR INDEX